Safety in the Laboratory

The laboratory can be an interesting and exciting place. However, the myriad of new equipment and materials that make the lab interesting and exciting can also produce hazards. Thus, an important objective of every laboratory course is to provide each student with the knowledge and techniques necessary to use the materials and equipment found in the lab safely. Failure to follow laboratory safety rules can lead to injury or death. Take all safety information seriously, seek a full understanding of the rules, and follow them throughout the time spent in the lab.

This sheet is an introduction to the concept of laboratory safety. An attempt to summarize all the information necessary for lab safety here would be incomplete and ineffective. Each experiment, chemical, or piece of equipment will present a new hazard on some level. Make sure you understand the hazards and precautions necessary whenever using a piece of equipment or chemical for the first time. Always review the safety rules when using a common device or material.

CHEMICAL HAZARDS

The Material Safety Data Sheet (MSDS) provides a summary of the hazards known about a chemical. Learn where the MSDS sheets are kept and consult them whenever using a chemical for the first time. It is important to follow all appropriate procedures accurately. The hazards posed by a chemical may be measured as the product of toxicity and the dose. Good lab procedures generally keep the "dose" at zero and consequently the actual hazard is usually small. Carefully following all appropriate laboratory practices will help ensure this.

SHARPS

The most common laboratory accidents involve cuts and punctures from broken glass or other "sharps" such as syringe needles. Be alert when using any "sharp" and use only the proper handling technique. Implement each technique carefully.

(please see reverse side for more safety tips)

To access electronic Periodic Tables as well as other valuable Science resources, visit Jones and Bartlett Publishers on the web. www.jbpub.com

W9-CPA-641

EYE PROTECTION

The human eye is an irreplaceable piece of equipment. Make sure everyone is properly wearing the approved eyewear for the laboratory whenever an operation that presents a chemical splash hazard is being performed. Wearing the best pair of goggles on the forehead or around the neck is obviously ineffective eye protection. Be sure your approved eye protection fits correctly.

PROPER ATTIRE

Proper attire is essential in the lab. There is always a chance of contacting a chemical spill or a sharp edge on a piece of equipment. An old pair of jeans (with few, if any, holes) and a shirt that covers from the waist to the shoulders and down to the biceps is generally the safest lab attire. Long hair, neckties, and loose fitting clothes present hazards and must always be restrained in the lab.

It is very important to wear shoes that cover the bottoms and the tops of your feet, because the lab floor is rarely completely free from debris, broken glass, or small chemical splashes. An old pair of walking shoes or sneakers that have no holes is the best footwear for lab operations. Sandals, open toe shoes, thongs, and high heels are not permitted in the lab.

GENERAL PROTOCOLS

In order to minimize contact with chemicals in the lab atmosphere, eating, drinking, and applying cosmetics is not permitted. Smoking and chewing tobacco products are also prohibited. Washing your hands before leaving the lab each day is recommended to minimize the dangers of chemical contact.

SAFETY EQUIPMENT

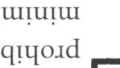

Every laboratory is equipped with several pieces of safety equipment. Know the locations of the eye wash, shower, fire extinguisher and other safety devices, as well as the procedures for their proper use. Be familiar with the evacuation procedures to be used in case of an emergency.

Each lab should also contain at least one copy of SAFETY in Academic Chemistry Laboratories, published by the American Chemical Society. (A free copy may be obtained by calling 1-800-227-5558.)

This sheet is not a substitute for proper and extensive safety training according to the Chemical Hygiene Plan implemented for the instructional laboratory in which you are working. Refer to the CHP whenever you have a question about safety or standard operating procedures.

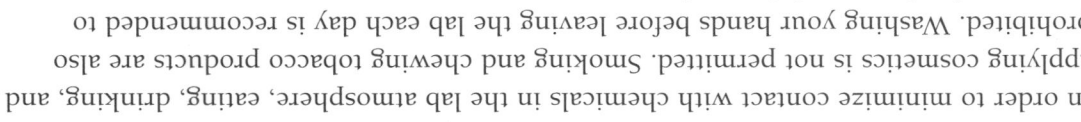

DATE	EXP. NUMBER	EXPERIMENT		01
NAME		LAB PARTNER		WITNESS

DATE	EXP. NUMBER	EXPERIMENT		01
NAME			LAB PARTNER	WITNESS

DATE	EXP. NUMBER	EXPERIMENT			
NAME			LAB PARTNER		WITNESS

DATE	EXP. NUMBER	EXPERIMENT		02
NAME		LAB PARTNER	WITNESS	

DATE	EXP. NUMBER	EXPERIMENT			03
NAME			LAB PARTNER		WITNESS

DATE	EXP. NUMBER	EXPERIMENT			03
NAME			LAB PARTNER		WITNESS

DATE	EXP. NUMBER	EXPERIMENT
NAME		LAB PARTNER	WITNESS

DATE	EXP. NUMBER	EXPERIMENT		04
NAME			LAB PARTNER	WITNESS

DATE	EXP. NUMBER	EXPERIMENT		05
NAME		LAB PARTNER	WITNESS	

DATE	EXP. NUMBER	EXPERIMENT		05
NAME			LAB PARTNER	WITNESS

DATE	EXP. NUMBER	EXPERIMENT		
NAME			LAB PARTNER	WITNESS

DATE	EXP. NUMBER	EXPERIMENT	06
NAME		LAB PARTNER	WITNESS

DATE	EXP. NUMBER	EXPERIMENT

NAME	LAB PARTNER	WITNESS

DATE	EXP. NUMBER	EXPERIMENT		
NAME			LAB PARTNER	WITNESS

DATE	EXP. NUMBER	EXPERIMENT		08
NAME		LAB PARTNER	WITNESS	

DATE	EXP. NUMBER	EXPERIMENT	
NAME		LAB PARTNER	WITNESS

DATE	EXP. NUMBER	EXPERIMENT		
NAME			LAB PARTNER	WITNESS

DATE	EXP. NUMBER	EXPERIMENT	
NAME		LAB PARTNER	WITNESS

DATE	EXP. NUMBER	EXPERIMENT	10
NAME		LAB PARTNER	WITNESS

DATE	EXP. NUMBER	EXPERIMENT		
NAME			LAB PARTNER	WITNESS

DATE	EXP. NUMBER	EXPERIMENT		WITNESS
NAME		LAB PARTNER		

DATE	EXP. NUMBER	EXPERIMENT		11
NAME		LAB PARTNER	WITNESS	

DATE	EXP. NUMBER	EXPERIMENT		
NAME			LAB PARTNER	WITNESS

DATE	EXP. NUMBER	EXPERIMENT		12
NAME		LAB PARTNER	WITNESS	

DATE	EXP. NUMBER	EXPERIMENT		
NAME			LAB PARTNER	WITNESS

LAB PARTNER

DATE	EXP. NUMBER	EXPERIMENT		
NAME			LAB PARTNER	WITNESS

DATE	EXP. NUMBER	EXPERIMENT		
NAME			LAB PARTNER	WITNESS

DATE	EXP. NUMBER	EXPERIMENT	
NAME		LAB PARTNER	WITNESS

DATE	EXP. NUMBER	EXPERIMENT		
NAME		LAB PARTNER		WITNESS

DATE	EXP. NUMBER	EXPERIMENT		16
NAME		LAB PARTNER		WITNESS

DATE	EXP. NUMBER	EXPERIMENT	
NAME		LAB PARTNER	WITNESS

DATE	EXP. NUMBER	EXPERIMENT			17
NAME			LAB PARTNER		WITNESS

DATE	EXP. NUMBER	EXPERIMENT		
NAME			LAB PARTNER	WITNESS

DATE	EXP. NUMBER	EXPERIMENT		
NAME			LAB PARTNER	WITNESS

DATE	EXP. NUMBER	EXPERIMENT		18
NAME			LAB PARTNER	WITNESS

DATE	EXP. NUMBER	EXPERIMENT		19
NAME			LAB PARTNER	WITNESS

DATE	EXP. NUMBER	EXPERIMENT		
NAME			LAB PARTNER	WITNESS

DATE	EXP. NUMBER	EXPERIMENT			
NAME			LAB PARTNER		WITNESS

DATE	EXP. NUMBER	EXPERIMENT			20
NAME			LAB PARTNER		WITNESS

DATE	EXP. NUMBER	EXPERIMENT		
NAME			LAB PARTNER	WITNESS

DATE	EXP. NUMBER	EXPERIMENT			
NAME			LAB PARTNER		WITNESS

DATE	EXP. NUMBER	EXPERIMENT		
NAME		LAB PARTNER		WITNESS

DATE	EXP. NUMBER	EXPERIMENT			22
NAME			LAB PARTNER	WITNESS	

DATE	EXP. NUMBER	EXPERIMENT	
NAME		LAB PARTNER	WITNESS

DATE	EXP. NUMBER	EXPERIMENT		
NAME			LAB PARTNER	WITNESS

DATE	EXP. NUMBER	EXPERIMENT			
NAME			LAB PARTNER		WITNESS

24

DATE	EXP. NUMBER	EXPERIMENT
NAME		LAB PARTNER

WITNESS

DATE	EXP. NUMBER	EXPERIMENT	
NAME		LAB PARTNER	WITNESS

DATE	EXP. NUMBER	EXPERIMENT		25
NAME			LAB PARTNER	WITNESS

DATE	EXP. NUMBER	EXPERIMENT

NAME	LAB PARTNER	WITNESS

DATE	EXP. NUMBER	EXPERIMENT
NAME		LAB PARTNER	WITNESS

DATE	EXP. NUMBER	EXPERIMENT		
NAME			LAB PARTNER	WITNESS

DATE	EXP. NUMBER	EXPERIMENT		
NAME			LAB PARTNER	WITNESS

DATE	EXP. NUMBER	EXPERIMENT		
NAME			LAB PARTNER	WITNESS

DATE	EXP. NUMBER	EXPERIMENT		
NAME			LAB PARTNER	WITNESS

DATE	EXP. NUMBER	EXPERIMENT		
NAME			LAB PARTNER	WITNESS

DATE	EXP. NUMBER	EXPERIMENT	
NAME		LAB PARTNER	WITNESS

DATE	EXP. NUMBER	EXPERIMENT		
NAME			LAB PARTNER	WITNESS

DATE	EXP. NUMBER	EXPERIMENT		
NAME			LAB PARTNER	WITNESS

DATE	EXP. NUMBER	EXPERIMENT		
NAME			LAB PARTNER	WITNESS

DATE	EXP. NUMBER	EXPERIMENT	
NAME		LAB PARTNER	WITNESS

DATE	EXP. NUMBER	EXPERIMENT
NAME		LAB PARTNER

WITNESS

DATE	EXP. NUMBER	EXPERIMENT					32
NAME			LAB PARTNER			WITNESS	

DATE	EXP. NUMBER	EXPERIMENT		
NAME			LAB PARTNER	WITNESS

DATE	EXP. NUMBER	EXPERIMENT
NAME		LAB PARTNER	WITNESS

DATE	EXP. NUMBER	EXPERIMENT		34
NAME		LAB PARTNER	WITNESS	

DATE	EXP. NUMBER	EXPERIMENT			
NAME			LAB PARTNER		WITNESS

34

DATE	EXP. NUMBER	EXPERIMENT		
NAME			LAB PARTNER	WITNESS

DATE	EXP. NUMBER	EXPERIMENT	
NAME		LAB PARTNER	WITNESS

DATE	EXP. NUMBER	EXPERIMENT	
NAME		LAB PARTNER	WITNESS

DATE	EXP. NUMBER	EXPERIMENT		36
NAME			LAB PARTNER	WITNESS

DATE	EXP. NUMBER	EXPERIMENT	
NAME		LAB PARTNER	WITNESS

DATE	EXP. NUMBER	EXPERIMENT		37
NAME		LAB PARTNER	WITNESS	

DATE	EXP. NUMBER	EXPERIMENT		
NAME			LAB PARTNER	WITNESS

DATE	EXP. NUMBER	EXPERIMENT		38
NAME			LAB PARTNER	WITNESS

DATE	EXP. NUMBER	EXPERIMENT		
NAME			LAB PARTNER	WITNESS

DATE	EXP. NUMBER	EXPERIMENT		
NAME			LAB PARTNER	WITNESS

DATE	EXP. NUMBER	EXPERIMENT

NAME	LAB PARTNER	WITNESS

DATE	EXP. NUMBER	EXPERIMENT	40
NAME		LAB PARTNER	WITNESS

DATE	EXP. NUMBER	EXPERIMENT	
NAME		LAB PARTNER	WITNESS

DATE	EXP. NUMBER	EXPERIMENT		41
NAME			LAB PARTNER	WITNESS

DATE	EXP. NUMBER	EXPERIMENT		42
NAME		LAB PARTNER	WITNESS

DATE	EXP. NUMBER	EXPERIMENT		42
NAME			LAB PARTNER	WITNESS

DATE	EXP. NUMBER	EXPERIMENT		WITNESS
NAME			LAB PARTNER	

DATE	EXP. NUMBER	EXPERIMENT		
NAME			LAB PARTNER	WITNESS

DATE	EXP. NUMBER	EXPERIMENT		
NAME		LAB PARTNER	WITNESS	

DATE	EXP. NUMBER	EXPERIMENT		WITNESS
NAME			LAB PARTNER	

DATE	EXP. NUMBER	EXPERIMENT	
NAME		LAB PARTNER	WITNESS

DATE	EXP. NUMBER	EXPERIMENT		45
NAME			LAB PARTNER	WITNESS

DATE	EXP. NUMBER	EXPERIMENT		46
NAME			LAB PARTNER	WITNESS

DATE	EXP. NUMBER	EXPERIMENT		WITNESS
NAME			LAB PARTNER	WITNESS

46

DATE	EXP. NUMBER	EXPERIMENT		47
NAME		LAB PARTNER	WITNESS	

| DATE | EXP. NUMBER | EXPERIMENT | | | 47 |
| NAME | | | LAB PARTNER | WITNESS | |

DATE	EXP. NUMBER	EXPERIMENT	
NAME		LAB PARTNER	WITNESS

DATE	EXP. NUMBER	EXPERIMENT		48
NAME		LAB PARTNER		WITNESS

DATE	EXP. NUMBER	EXPERIMENT		
NAME			LAB PARTNER	WITNESS

DATE	EXP. NUMBER	EXPERIMENT		49
NAME			LAB PARTNER	WITNESS

DATE	EXP. NUMBER	EXPERIMENT		WITNESS
NAME			LAB PARTNER	

DATE	EXP. NUMBER	EXPERIMENT		
NAME			LAB PARTNER	WITNESS

DATE	EXP. NUMBER	EXPERIMENT		
NAME			LAB PARTNER	WITNESS

DATE	EXP. NUMBER	EXPERIMENT	
NAME		LAB PARTNER	WITNESS

DATE	EXP. NUMBER	EXPERIMENT		
NAME			LAB PARTNER	WITNESS

DATE	EXP. NUMBER	EXPERIMENT		52
NAME			LAB PARTNER	WITNESS

DATE	EXP. NUMBER	EXPERIMENT	
NAME		LAB PARTNER	WITNESS

DATE	EXP. NUMBER	EXPERIMENT		
NAME			LAB PARTNER	WITNESS

DATE	EXP. NUMBER	EXPERIMENT		WITNESS
NAME			LAB PARTNER	

DATE	EXP. NUMBER	EXPERIMENT	
NAME		LAB PARTNER	WITNESS

DATE	EXP. NUMBER	EXPERIMENT		
NAME			LAB PARTNER	WITNESS

DATE	EXP. NUMBER	EXPERIMENT	
NAME		LAB PARTNER	WITNESS

DATE	EXP. NUMBER	EXPERIMENT		
NAME			LAB PARTNER	WITNESS

DATE	EXP. NUMBER	EXPERIMENT		
NAME			LAB PARTNER	WITNESS

DATE	EXP. NUMBER	EXPERIMENT		
NAME			LAB PARTNER	WITNESS

DATE	EXP. NUMBER	EXPERIMENT		57
NAME			LAB PARTNER	WITNESS

DATE	EXP. NUMBER	EXPERIMENT		58
NAME		LAB PARTNER	WITNESS	

DATE	EXP. NUMBER	EXPERIMENT	
NAME		LAB PARTNER	WITNESS

DATE	EXP. NUMBER	EXPERIMENT			59
NAME			LAB PARTNER	WITNESS	

DATE	EXP. NUMBER	EXPERIMENT	59
NAME		LAB PARTNER	WITNESS

DATE	EXP. NUMBER	EXPERIMENT		WITNESS
NAME		LAB PARTNER		

DATE	EXP. NUMBER	EXPERIMENT		
NAME			LAB PARTNER	WITNESS

DATE	EXP. NUMBER	EXPERIMENT		
NAME			LAB PARTNER	WITNESS

DATE	EXP. NUMBER	EXPERIMENT	
NAME		LAB PARTNER	WITNESS

DATE	EXP. NUMBER	EXPERIMENT		
NAME		LAB PARTNER		WITNESS

DATE	EXP. NUMBER	EXPERIMENT	
NAME		LAB PARTNER	WITNESS

DATE	EXP. NUMBER	EXPERIMENT		WITNESS
NAME		LAB PARTNER		

DATE	EXP. NUMBER	EXPERIMENT			
NAME			LAB PARTNER		WITNESS

DATE	EXP. NUMBER	EXPERIMENT		
NAME			LAB PARTNER	WITNESS

DATE	EXP. NUMBER	EXPERIMENT		
NAME			LAB PARTNER	WITNESS

DATE	EXP. NUMBER	EXPERIMENT	
NAME		LAB PARTNER	WITNESS

DATE	EXP. NUMBER	EXPERIMENT		
NAME			LAB PARTNER	WITNESS

DATE	EXP. NUMBER	EXPERIMENT		
NAME			LAB PARTNER	WITNESS

DATE	EXP. NUMBER	EXPERIMENT		NAME		LAB PARTNER	WITNESS

DATE	EXP. NUMBER	EXPERIMENT
NAME		LAB PARTNER	WITNESS

DATE	EXP. NUMBER	EXPERIMENT		
NAME			LAB PARTNER	WITNESS

DATE	EXP. NUMBER	EXPERIMENT	
NAME		LAB PARTNER	WITNESS

DATE	EXP. NUMBER	EXPERIMENT
NAME		LAB PARTNER

WITNESS

DATE	EXP. NUMBER	EXPERIMENT	
NAME		LAB PARTNER	WITNESS

DATE	EXP. NUMBER	EXPERIMENT		
NAME			LAB PARTNER	WITNESS

DATE	EXP. NUMBER	EXPERIMENT	
NAME		LAB PARTNER	WITNESS

DATE	EXP. NUMBER	EXPERIMENT		
NAME			LAB PARTNER	WITNESS

DATE	EXP. NUMBER	EXPERIMENT
NAME		LAB PARTNER	WITNESS

DATE	EXP. NUMBER	EXPERIMENT
NAME		LAB PARTNER	WITNESS

DATE	EXP. NUMBER	EXPERIMENT		
NAME			LAB PARTNER	WITNESS

DATE	EXP. NUMBER	EXPERIMENT	
NAME		LAB PARTNER	WITNESS

DATE	EXP. NUMBER	EXPERIMENT		
NAME		LAB PARTNER		WITNESS

73

DATE	EXP. NUMBER	EXPERIMENT		73
NAME		LAB PARTNER	WITNESS	

DATE	EXP. NUMBER	EXPERIMENT	
NAME		LAB PARTNER	WITNESS

DATE	EXP. NUMBER	EXPERIMENT

NAME	LAB PARTNER	WITNESS

DATE	EXP. NUMBER	EXPERIMENT		
NAME			LAB PARTNER	WITNESS

DATE	EXP. NUMBER	EXPERIMENT		
NAME			LAB PARTNER	WITNESS

DATE	EXP. NUMBER	EXPERIMENT		
NAME			LAB PARTNER	WITNESS

DATE	EXP. NUMBER	EXPERIMENT		
NAME			LAB PARTNER	WITNESS

DATE	EXP. NUMBER	EXPERIMENT			
NAME			LAB PARTNER		WITNESS

DATE	EXP. NUMBER	EXPERIMENT	
NAME		LAB PARTNER	WITNESS

DATE	EXP. NUMBER	EXPERIMENT

NAME	LAB PARTNER	WITNESS

DATE	EXP. NUMBER	EXPERIMENT
NAME		LAB PARTNER	WITNESS

DATE	EXP. NUMBER	EXPERIMENT		
NAME			LAB PARTNER	WITNESS

DATE	EXP. NUMBER	EXPERIMENT	
NAME		LAB PARTNER	WITNESS

DATE	EXP. NUMBER	EXPERIMENT
NAME		LAB PARTNER	WITNESS

DATE	EXP. NUMBER	EXPERIMENT
NAME		LAB PARTNER	WITNESS

DATE	EXP. NUMBER	EXPERIMENT
NAME		LAB PARTNER	WITNESS

DATE	EXP. NUMBER	EXPERIMENT			
NAME			LAB PARTNER		WITNESS

DATE	EXP. NUMBER	EXPERIMENT		
NAME			LAB PARTNER	WITNESS

DATE	EXP. NUMBER	EXPERIMENT		
NAME			LAB PARTNER	WITNESS

DATE	EXP. NUMBER	EXPERIMENT

NAME	LAB PARTNER	WITNESS

DATE	EXP. NUMBER	EXPERIMENT			83
NAME			LAB PARTNER		WITNESS

DATE	EXP. NUMBER	EXPERIMENT		
NAME			LAB PARTNER	WITNESS

DATE	EXP. NUMBER	EXPERIMENT	
NAME		LAB PARTNER	WITNESS

DATE	EXP. NUMBER	EXPERIMENT		
NAME		LAB PARTNER		WITNESS

DATE	EXP. NUMBER	EXPERIMENT		85
NAME		LAB PARTNER		WITNESS

DATE	EXP. NUMBER	EXPERIMENT		
NAME			LAB PARTNER	WITNESS

DATE	EXP. NUMBER	EXPERIMENT	
NAME		LAB PARTNER	WITNESS

DATE	EXP. NUMBER	EXPERIMENT

NAME	LAB PARTNER	WITNESS

DATE	EXP. NUMBER	EXPERIMENT
NAME		LAB PARTNER

WITNESS

DATE	EXP. NUMBER	EXPERIMENT		
NAME		LAB PARTNER	WITNESS	

DATE	EXP. NUMBER	EXPERIMENT		
NAME			LAB PARTNER	WITNESS

DATE	EXP. NUMBER	EXPERIMENT		WITNESS
NAME		LAB PARTNER		

DATE	EXP. NUMBER	EXPERIMENT			89
NAME			LAB PARTNER	WITNESS	

DATE	EXP. NUMBER	EXPERIMENT
NAME		LAB PARTNER	WITNESS

DATE	EXP. NUMBER	EXPERIMENT	
NAME		LAB PARTNER	WITNESS

DATE	EXP. NUMBER	EXPERIMENT		
NAME		LAB PARTNER		WITNESS

DATE	EXP. NUMBER	EXPERIMENT		
NAME			LAB PARTNER	WITNESS

DATE	EXP. NUMBER	EXPERIMENT	
NAME		LAB PARTNER	WITNESS

DATE	EXP. NUMBER	EXPERIMENT	
NAME		LAB PARTNER	WITNESS

DATE	EXP. NUMBER	EXPERIMENT		
NAME			LAB PARTNER	WITNESS

DATE	EXP. NUMBER	EXPERIMENT		93
NAME		LAB PARTNER	WITNESS	

DATE	EXP. NUMBER	EXPERIMENT	
NAME		LAB PARTNER	WITNESS

DATE	EXP. NUMBER	EXPERIMENT
NAME		LAB PARTNER	WITNESS

DATE	EXP. NUMBER	EXPERIMENT		95
NAME		LAB PARTNER	WITNESS	

DATE	EXP. NUMBER	EXPERIMENT		WITNESS
NAME			LAB PARTNER	

DATE	EXP. NUMBER	EXPERIMENT		
NAME			LAB PARTNER	WITNESS

DATE	EXP. NUMBER	EXPERIMENT		
NAME			LAB PARTNER	WITNESS

DATE	EXP. NUMBER	EXPERIMENT

NAME	LAB PARTNER	WITNESS

DATE	EXP. NUMBER	EXPERIMENT	
NAME		LAB PARTNER	WITNESS

DATE	EXP. NUMBER	EXPERIMENT	
NAME		LAB PARTNER	WITNESS

DATE	EXP. NUMBER	EXPERIMENT		98
NAME			LAB PARTNER	WITNESS

DATE	EXP. NUMBER	EXPERIMENT

NAME	LAB PARTNER	WITNESS

DATE	EXP. NUMBER	EXPERIMENT		
NAME			LAB PARTNER	WITNESS